THE
LITTLE PURPLE PROBABILITY
BOOK

一本小小的紫色概率书

（加）布兰登·罗伊尔◎著
周丽萍◎译

图书在版编目（CIP）数据

一本小小的紫色概率书 /（加）布兰登·罗伊尔著；周丽萍译 . -- 北京：九州出版社，2017.4

ISBN 978-7-5108-5127-8

Ⅰ . ①一… Ⅱ . ①布… ②周… Ⅲ . ①概率－普及读物 Ⅳ . ① O211.1-49

中国版本图书馆 CIP 数据核字 (2017) 第 055777 号

The Little Purple Probability Book
Copyright © Brandon Royal 2016
All rights reserved. No portion of this book may be reproduced, stored in a retrieval system, or transmitted in any form or by any means—electronic, mechanical, photocopy, recording, scanning, or other—except for brief quotations in critical reviews or articles, without the prior written permission of the publisher.
Simplified edition is arranged through CA-LINK International LLC(www.ca-link.com)

版权合同登记号 图字：01-2016-9690

一本小小的紫色概率书

作　　者	（加）布兰登·罗伊尔 著　周丽萍 译
出版发行	九州出版社
地　　址	北京市西城区阜外大街甲 35 号（100037）
发行电话	（010）68992190/3/5/6
网　　址	www.jiuzhoupress.com
电子信箱	jiuzhou@jiuzhoupress.com
印　　刷	三河市华成印务有限公司
开　　本	710 毫米 ×930 毫米　16 开
印　　张	5.5
字　　数	48 千字
版　　次	2017 年 5 月第 1 版
印　　次	2017 年 5 月第 1 次印刷
书　　号	ISBN 978-7-5108-5127-8
定　　价	38.00 元

★版权所有　侵权必究★

THE
LITTLE PURPLE
PROBABILITY
BOOK

目录

前言	001
概述	003
基本概率公式	001
附加公式	017
经典题型：概率	023
经典题型：枚举	031
经典题型：排列	033
经典题型：组合	039
答案和解析	043

THE
LITTLE PURPLE
PROBABILITY
BOOK

前言

编撰这本小书，主要是为了帮助读者在尽可能短的时间内掌握基础概率。

达到这个目标需要两个步骤。

第一步是制定便于理解概率的宏观概念框架。本书前部分的概述章节含有两个战略流程图——一个是概率，另一个是排列组合——这两个是本书要讨论的重点主题。书中一共归纳总结了14个基本公式，每个公式附相关习题。

第二步是练习少量最经典的概率题，它们代表了可能出现在其他地方的较普遍的概率题型。

掌握基本概率问题的思维技巧，要求具备将给定题目与正确公式相匹配的能力。要将题目恰当归类，理解容易混淆的有关概率及相关主题的术语很有必要。例如，两个事件独立或两个事件非互斥是什么意思？我们自信能够将问题归类为常见类型时，就能判断某个具体题目是否属于某个经典题型的变体，或者它是否包含常见的容易忽视的陷阱。书中插入了特别提示（标注了"注意"）。这些附加信息、例子或补充题目，无疑大有裨益。

这本概率书阐述了涉及算术和代数基础的、有一定难度的数学题。概率这门课程需要特别关注，因此独立成书。

让我们开始吧。

THE
LITTLE PURPLE
PROBABILITY
BOOK

概述

　　本书共包含三大主题：概率、排列、组合。那么，概率、排列、组合之间有什么区别呢？

　　概率是介于 0 和 1 之间的小数或分数（1 是必然概率，0 是不可能概率）。换句话说，概率是介于 0% 和 100% 之间的百分数（100% 是必然概率，0% 是不可能概率）。而排列和组合则是大于或等于 1 的结果，通常它们的结果相当大，如 10、36、720 等。

概率：

关于概率，一个简易的经验法则是，先判断面对的是"和"情形还是"或"情形。"和"意味着相乘，"或"意味着相加。例如，如果题目问"x 和 y 的概率是多少"，我们就把两个单独的概率相乘。如果题目问"x 或 y 的概率是多少"，我们就把两个单独的概率相加。

另外，如果一道概率题要求相乘，则还需要进一步问："它们是独立事件还是相关事件？"独立是指两个事件互不影响，只需将两个单独的概率简单相乘得出最后答案。不独立（相关）是指其中一个事件的发生会影响另一个事件的发生，这种影响必须考虑进去。

同样，如果一道概率题要求相加，也需要进一步问："它们是互斥事件还是非互斥事件？"互斥是指两个事件不可能同时发生，不会出现"重复"。如果两个事件互斥，直接将概率相加。非互斥是指两个事件可能同时发生，因而出现重复。如果两个事件非互斥，则重复部分不能计算两次。

排列和组合：

关于排列和组合，排列是有序集合，组合是无序集合。顺序影响排列，但不影响组合。例如，在排列中，*AB* 和 *BA* 是两种不同的结果，但在组合中它们是一个结果。现实生活中，排列的例子有电话号码、汽车牌照、邮编和密码。组合的例子有选择组员或彩票。以彩票为例，数字的顺序没有影响，我们只需要集齐所有数字即可，通常六个就行。

在实际应用中，通常"安排"或"可能性"等词语暗示排列，"选取"或"选择"等词语暗示组合。

阶乘：

阶乘是指用这样的方式进行乘法运算：

例：$4! = 4 \times 3 \times 2 \times 1$

例：$7! = 7 \times 6 \times 5 \times 4 \times 3 \times 2 \times 1$

0 的阶乘等于 1，1 的阶乘也等于 1：

例：$0! = 1$

例：$1! = 1$

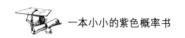一本小小的紫色概率书

硬币、骰子、弹珠和纸牌：

这本概率书中的题目用到了硬币、骰子、弹珠和纸牌。具体如下：

硬币的两面分别是正面和反面。

骰子有六面，数字为1~6，每次掷出后每一面出现的可能性相同。

弹珠假设为单一素色。

一副纸牌有52张，平均分为四个花色——梅花、方片、红桃、黑桃——每种花色13张牌，包括A、K、Q、J、10、9、8、7、6、5、4、3、2。

图表A 概率流程图

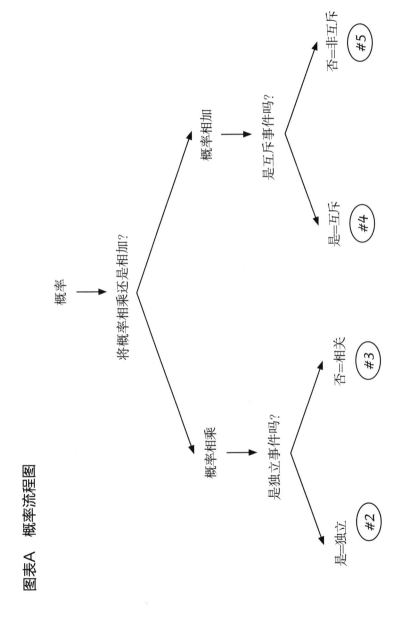

上面的数字表示适用的概率公式。

图表B 排列和组合流程图

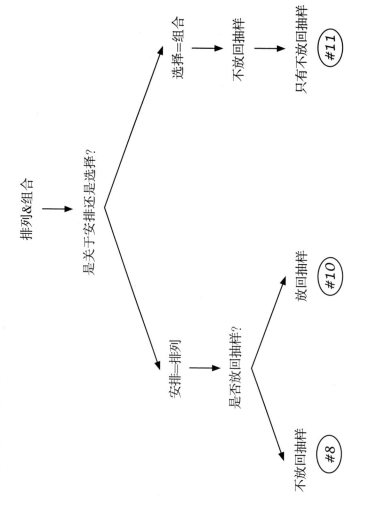

上面的数字表示适用的排列或组合公式。

THE
LITTLE PURPLE
PROBABILITY
BOOK

基本概率公式

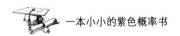

通用公式

#1 概率 = 被选择事件 / 总可能性数量

例：一共卖出 10000 张抽奖券，你买了 3 张，只有一张有奖，那么你中奖的概率是多少？

概率 = $\dfrac{3}{10000}$

特殊乘法法则

#2 $P(AB) = P(A) \times P(B)$

> A 和 B 的概率等于 A 的概率乘以 B 的概率。

如果事件相互独立（即一个事件不影响另一事件），只需将它们各自的概率相乘。

例：投掷硬币两次，第一次和第二次都是正面朝上的概率是多少？

$$\frac{1}{2} \times \frac{1}{2} = \frac{1}{4}$$

通用乘法法则

#3 $P(AB) = P(A) \times P(B/A)$

> 假设事件 A 已经发生，A 和 B 的概率等于 A 的概率乘以 B 的概率。

如果事件不独立（即一个事件会影响另一个事件），我们必须在第一个事件造成的影响的基础上，调整第二个事件。

例：一个袋子里装了 6 颗弹珠，3 颗蓝色，3 颗绿色。先后从包里取出 2 颗绿色弹珠的概率是多少？

$$\frac{3}{6} \times \frac{2}{5} = \frac{6}{30} = \frac{1}{5}$$

特殊加法法则

#4 $P(A+B) = P(A) + P(B)$

A 或 B 的概率等于 A 的概率加 B 的概率。

如果事件互斥（即不重复），只需将事件的概率相加即可。

例：山姆去蒙特卡姆学院中学的概率是 50%，去克莱森特高地中学的概率是 25%。那么，山姆去蒙特卡姆学院中学或克莱森特高地中学的概率是多少？

$$50\% + 25\% = 75\%$$

通用加法法则

#5 $P(A+B) = P(A) + P(B) - P(AB)$

> A 或 B 的概率等于 A 的概率加 B 的概率，再减去 A 和 B 的概率。

如果事件不互斥（即存在重复），我们将分别的概率相加之后还要减去重复部分。

例：明天下雨的概率是 30%，刮风的概率是 20%。那么明天下雨或刮风的概率是多少？

$$30\% + 20\% - (30\% \times 20\%)$$
$$= 30\% + 20\% - 6\%$$
$$= 50\% - 6\%$$
$$= 44\%$$

关于通用加法法则，减去重复部分是因为不能计算两次。两个事件重复时，肯定包含相同内容。因此，必须减去一次，避免"重复"计算。

> **注 意**
>
> 我们来快速比较一下通常所说的包含性"或"和排他性"或"。
>
> 证明概率公式#5所用的题目受包含性"或"的限制。我们有理由假设明天的天气既刮风又下雨。题目实际上问的是："明天的天气下雨或刮风或既下雨又刮风的概率是多少？"当事件存在重复时就会出现包含性"或"。
>
> 证明概率公式#4所用的题目，实际上问的是："山姆去蒙特卡姆学院中学或克莱森特高地中学，但不同时去两所高中的概率是多少？"在两个地点不同的高中之间所做的选择是互斥的，因此处理这道特定问题涉及排他性"或"。

互补法则

#6 $P(A) = 1 - P(\bar{A})$

A 发生的概率等于 1 减去 A 不发生的概率。

概率的互补法则描述的是概率的相减或相除，而不是概率的相加或相乘。用这个法则计算一个事件的概率，我们可以计算某个给定事件不发生的概率，然后用 1 减去该结果。

例：掷一对骰子，不掷出两个 6 的概率是多少？

掷出两个 6 的概率是：

$$\frac{1}{6} \times \frac{1}{6} = \frac{1}{36}$$

不掷出两个 6 的概率是：

$$1-\frac{1}{36}=\frac{35}{36}$$

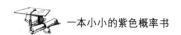

枚举法则

#7 如果做第一件事有 X 种方法，做第二件事有 Y 种方法，做第三件事有 Z 种方法，那么做这些事情的总方法数为 $x \times y \times z$。这被认为是枚举的一般法则。（在进行归纳推理时，如果逐个考察了某类事件的所有可能情况，从而得出一般结论，那么这个结论是可靠的，这种归纳方法叫作枚举法。——编者注）

> **注 意**
>
> 严格来讲，枚举法则不在"概率""排列"或"组合"的范畴。但是，在实际运用中常放在一起讨论。

例：速宴餐厅为顾客提供了一份菜单，每一类可选一份：2种不同的沙拉，3种不同的汤，5种不同的主菜，3种不同的甜点，以及咖啡或茶。顾客选餐的方式有多少种可能性？

$2\times3\times5\times3\times2=180$（种）

排列

#8 不放回抽样 $P_n^r = \dfrac{n!}{(n-r)!}$

$n=$ 项目总数，$r=$ 选取或排列的项目数

例：假设同一字母在任何特定编码中出现不超过一次，则 A、B、C、D 四个字母可以组成多少个双字母编码？

$$P_n^r = \dfrac{n!}{(n-r)!}$$

$$P_4^2 = \dfrac{4!}{(4-2)!} = \dfrac{4!}{2!} = \dfrac{4 \times 3 \times 2 \times 1}{2 \times 1} = 4 \times 3 = 12 \text{（个）}$$

#9 $P_n^n = n!$

> 所有项目全部选取时的快捷公式。

例：把4本不同的书摆放（或陈列）在书架上，有多少种摆法？

$$P_n^r = \frac{n!}{(n-r)!}$$

$$P_4^4 = \frac{4!}{(4-4)!} = \frac{4!}{0!} = \frac{4}{1} = 4! = 4 \times 3 \times 2 \times 1 = 24 \text{（种）}$$

或者，使用快捷公式：$n! = 4! = 24$（种）

#10 放回抽样 n^r

例：假设同一数字在任何特定编码中出现不超过一次，则数字1、2、3、4可以组成多少个四位数的编码？

$$4^4 = 256（个）$$

> **注意**
>
> 放回抽样的排列（即 n^r）严格来讲归属枚举法则范畴。为便于说明故包含在此。

组合

$$\boxed{\#11} \quad C_n^r = \frac{n!}{r!(n-r)!}$$

$n=$ 给出的项目总数，$r=$ 将要选择的项目数

例：为了粉刷房间内壁，从4种颜色中选择3种，有多少种选法？

$$C_n^r = \frac{n!}{r!(n-r)!}$$

$$C_4^3 = \frac{4!}{3!(4-3)!} = \frac{4!}{3!(1)!} = \frac{4 \times 3 \times 2 \times 1}{3 \times 2 \times 1 \times 1} = 4 \text{（种）}$$

THE
LITTLE PURPLE
PROBABILITY
BOOK

附 加 公 式

联合排列

#12 $\quad P_n^r \times P_n^r = \dfrac{n!}{(n-r)!} \times \dfrac{n!}{(n-r)!}$

例：一名游客计划去 5 个西欧城市中的 3 个城市旅游，然后去 4 个东欧城市中的 2 个城市旅游。这名游客可能有多少种路线？

$$P_5^3 \times P_4^2$$

$$= \dfrac{5!}{(5-3)!} \times \dfrac{4!}{(4-2)!}$$

$$= \dfrac{5!}{2!} \times \dfrac{4!}{2!}$$

$$= \dfrac{5\times 4\times 3\times 2\times 1}{2\times 1} \times \dfrac{4\times 3\times 2\times 1}{2\times 1}$$

$$=60\times12$$

$$=720（种）$$

将结果相乘,而不是相加,与枚举法则给出的解决方法一致。

联合组合

#13 $C_n^r \times C_n^r = \dfrac{n!}{r!(n-r)!} \times \dfrac{n!}{r!(n-r)!}$

例：要从 5 名专业高尔夫选手和 5 名专业网球选手中选出部分选手组成一个营销特别工作组。如果最后决定工作组是由 3 名高尔夫选手和 3 名网球选手组成，那么可能有多少种不同的工作组？

$$C_5^3 \times C_5^3$$

$$= \dfrac{5!}{3!(5-3)!} \times \dfrac{5!}{3!(5-3)!}$$

$$= \dfrac{5\times4\times3\times2\times1}{3\times2\times1\times2\times1} \times \dfrac{5\times4\times3\times2\times1}{3\times2\times1\times2\times1}$$

$$=10\times10=100\,(\text{种})$$

重复数字或字母（排列）

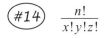 $\dfrac{n!}{x!y!z!}$

> x、y、z 指重复的字母或数字。

例：用四个数字 0、0、1、2 可以组成多少个四位数的编码？

$$\frac{4!}{2!}=\frac{4\times3\times2\times1}{2\times1}=12（个）$$

注意

2！指的是代表重复的两个数字 0。

THE
LITTLE PURPLE
PROBABILITY
BOOK

经典题型：概率

第1题　四张A

从一副纸牌中随机抽取4张，下面哪一项表示抽到4张A的概率？（纸牌逐一抽取，不放回）。

A）$\frac{1}{52} \times \frac{1}{52} \times \frac{1}{52} \times \frac{1}{52}$　　　　B）$\frac{1}{52} \times \frac{1}{51} \times \frac{1}{50} \times \frac{1}{49}$

C）$\frac{4}{52} \times \frac{3}{51} \times \frac{2}{50} \times \frac{1}{49}$　　　　D）$\frac{4}{52} \times \frac{3}{52} \times \frac{2}{52} \times \frac{1}{52}$

E）$\frac{4}{52} \times \frac{4}{52} \times \frac{4}{52} \times \frac{4}{52}$

经典题型：概率

第 2 题　橙色和蓝色（一）

一个袋子里有 5 颗弹珠——2 颗橙色，3 颗蓝色。如果从袋中取出 2 颗弹珠，第一颗是橙色且第二颗是蓝色的概率是多少？

A）$\dfrac{6}{25}$　　　　　　　　B）$\dfrac{3}{10}$

C）$\dfrac{2}{5}$　　　　　　　　D）$\dfrac{3}{5}$

E）$\dfrac{7}{10}$

第 3 题　橙色和蓝色（二）

一个袋子里有 5 颗弹珠——2 颗橙色，3 颗蓝色。如果从袋中取出 2 颗弹珠，至少一颗为橙色的概率是多少？

A）$\dfrac{6}{25}$　　　　　　　　B）$\dfrac{3}{10}$

C）$\dfrac{2}{5}$　　　　　　　　D）$\dfrac{3}{5}$

E）$\dfrac{7}{10}$

第4题 考试时间（一）

一名学生期末考试要考两门课程。他通过第一门课程的概率是 $\frac{3}{4}$，通过第二门课程的概率是 $\frac{2}{3}$。他通过第一门或第二门课程的概率是多少？

A) $\frac{5}{12}$

B) $\frac{1}{2}$

C) $\frac{7}{12}$

D) $\frac{5}{7}$

E) $\frac{11}{12}$

第 5 题　第六感

第一次掷骰子或第二次掷骰子，掷出 6 的概率是多少？

A) $\dfrac{1}{36}$　　　　　　　　B) $\dfrac{5}{18}$

C) $\dfrac{1}{6}$　　　　　　　　D) $\dfrac{11}{36}$

E) $\dfrac{1}{3}$

第6题　考试时间（二）

一名学生期末考试要考三门课程。他通过第一门课程的概率是 $\frac{3}{4}$，通过第二门课程的概率是 $\frac{2}{3}$，通过第三门课程的概率是 $\frac{1}{2}$。他通过至少一门课程的概率是多少？（或者：他通过第一或第二或第三门课程的概率是多少？）

A）$\frac{1}{4}$　　　　　　　　　B）$\frac{11}{24}$

C）$\frac{17}{24}$　　　　　　　　D）$\frac{3}{4}$

E）$\frac{23}{24}$

THE
LITTLE PURPLE
PROBABILITY
BOOK

经典题型：枚举

第 7 题　招聘

一家公司计划招聘一名销售经理、一名运务员和一名接待。公司已经缩小了候选人范围，计划面试剩下的候选人：7 名销售经理岗位候选人、4 名运务员岗位候选人和 10 名接待岗位候选人。这三个岗位的招聘有多少种可能？

A）$7 \times 4 \times 10$

B）$7 + 4 + 10$

C）$21 \times 20 \times 19$

D）$7! + 4! + 10!$

E）$7! \times 4! \times 10!$

THE
LITTLE PURPLE
PROBABILITY
BOOK

经典题型：排列

第8题 击剑

4名参赛人员分别代表4个不同的国家进入击剑冠军决赛。假设所有参赛者取胜的机会相等,则第一名和第二名奖牌的颁发有多少种可能?

A) 6   B) 7

C) 12   D) 16

E) 24

经典题型：排列

第9题　座次（一）

6名学生肩并肩坐成一排准备参加补考。他们有多少种坐法？

A）12　　　　　　　　　B）36

C）72　　　　　　　　　D）240

E）720

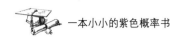

第10题 座次（二）

6名学生——3个男孩和3个女孩——肩并肩坐成一排准备参加补考。假设男孩与男孩不能相邻，且女孩与女孩不能相邻，那么他们有多少种坐法？

A）12

B）36

C）72

D）240

E）720

第11题 Banana

用单词 *BANANA* 的六个字母组成的编码有多少种方法？下面哪一项为正确的数学解法？

A）6!

B）$6!-(3!\times 2!)$

C）$6!-(3!+2!)$

D）$\dfrac{6!}{3!\times 2!}$

E）$\dfrac{6!}{3!+2!}$

第12题 餐桌

一张餐桌有5把椅子,3个人就座,剩2把空椅子,有多少种坐法?

A)8

B)12

C)60

D)118

E)120

经典题型：组合

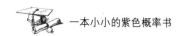

第13题 歌手

为了即将举行的慈善活动,一位男歌唱家要演唱6首"老歌"中的4首和5首"新歌"中的2首。这位歌手有多少种选择方式?

A)25

B)50

C)150

D)480

E)600

经典题型：组合

第14题 重逢

如果11个人重逢，每个人都与其他人一一握手，总握手次数是多少？

A）11×10×9×8×7×6×5×4×3×2×1

B）10×9×8×7×6×5×4×3×2×1

C）11×10

D）55

E）45

第15题 结果

假设 $P_n^r = \dfrac{n!}{(n-r)!}$ 且 $C_n^r = \dfrac{n!}{r!(n-r)!}$，$n$ 是项目总数，r 是被选取或选择的项目数，关于产生的结果数，下列哪些表述正确？

Ⅰ. $P_5^3 > P_5^2$ Ⅱ. $C_5^3 > C_5^2$ Ⅲ. $C_5^2 > P_5^2$

A）只有Ⅰ

B）只有Ⅰ和Ⅱ

C）只有Ⅰ和Ⅲ

D）只有Ⅱ和Ⅲ

E）Ⅰ和Ⅱ和Ⅲ

THE
LITTLE PURPLE
PROBABILITY
BOOK

答案和解析

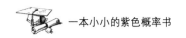

第1题　四张A

答案：C

解析：这道题目体现了概率通用乘法法则的典型运用。"不放回抽样"选择纸牌会影响下一张被选纸牌的概率，因为这一副纸牌中已经没有第一张了。"放回抽样"选择纸牌不会影响下一张被选纸牌的概率，因为放回已选纸牌使这副纸牌恢复原样。

第一次抽取，抽中 A 的机率是 $\frac{4}{52}$。第二次抽取，抽中 A 的机率是 $\frac{3}{51}$，因为可选的纸牌 A 少了一张，总纸牌数少了一张。第三次抽取，抽中 A 的机率是 $\frac{2}{50}$，因为可选的纸牌 A 又少了一张，总纸牌数也少了一张。最后一次抽取，抽中 A 的机率是 $\frac{1}{49}$，因为只剩一

张 A，而这副牌也刚好只剩 49 张。

顺便说一下，如果我们以放回抽样的方式抽取四张 A，则选项 E 为正确答案。例如："假如每抽完一张后放回再抽下一张，那么连续抽中四张 A 的概率是多少？"如果以放回抽样的方式抽取某张单独的牌，则选项 A 为正确答案。例如："如果从一副纸牌中随机抽取一张纸牌，每次抽完以后放回去，那么连续四次抽中黑桃 A 的概率是多少？"如果我们要按照特定顺序抽取四张 A，且每次所抽取的牌不放回，则选项 B 为正确答案。例如："先后抽中黑桃 A、红桃 A、梅花 A、方片 A 的概率是多少？"就是这种情况。

> **注 意**
>
> 学会如何分辨基本概率问题，有两组术语很关键。第一组是"互斥"和"非互斥"，第二组是"独立"和"非独立（相关）"。互斥是指两个事件或结果彼此不重复或不同时发生；非互斥是指两个事件彼此重复或可能同时发生。独立是指两个事件或结果互不影响，相对彼此而言随机发生；不独立（相关）是指两个事件或结果互相影响，一个事件的发生会影响另一个事件的发生。
>
> 以下是一些说明这些术语的实例。

假设我们要举行一个商务会议，准备邀请参会人员和发言嘉宾。VIP 座次和非 VIP 座次的安排是互斥结果。任何人要么在 VIP 席位，要么不在。同理，判断参会人员是本州人士还是非本州人士情况相同，因为参会人员要么来自本州，要么来自其他州，不可能出现重叠。但是，按职业对参会人员进行分类，谁是经理、谁是工程师、谁是销售人员、谁是企业家，这之间可能出现重叠，有些参会人员可能归属多个类别。因此，这些分类非互斥。

两项工作可能独立，对彼此无影响。例如，筹备会议时，先制名牌再复印会议材料或者先复印会议材料再制名牌并没有区别。这些事件代表彼此没有关联的单独任务。

两个事件也可能不独立，他们可能互相影响，必须按照特定顺序发生的事件就是这种情况。筹备会议时，我们邀请嘉宾作会议演讲前，必须先做会议计划。同样，参会人员必须先报名，之后才可以参会，然后才可以填写会后的会议评估表。换句话说，会议评估表的填写取决于一个人是否确实参加了会议，而反过来，一个人是否参会取决于他是否先报了名。

第2题 橙色和蓝色（一）

答案： B

解析： 这道题用通用乘法法则解答。题中使用了"且"这个词，提示我们需要把概率相乘。选中第一颗橙色弹珠的概率会影响选中第一颗蓝色弹珠的概率，因为备选弹珠少了一颗（不放回抽样选择）。

选中一颗橙色弹珠的概率是 $\frac{2}{5}$，选中一颗蓝色弹珠的概率是 $\frac{3}{4}$。

$$\frac{2}{5} \times \frac{3}{4} = \frac{6}{20} = \frac{3}{10}$$

注意陷阱选项 A，忘记从第二个分数的分母中减除一颗弹珠，计算成：

$$\frac{2}{5} \times \frac{3}{5} = \frac{6}{25}$$

第3题　橙色和蓝色（二）

答案： E

解析： 这道题运用了概率的互补法则。见概率公式#6（第008页）。

解答这道题最简单的方法是看我们不想要什么东西。"至少一颗橙色弹珠"是指我们除不要两颗蓝色弹珠以外其他都可以。

拿到两颗蓝色弹珠的概率的：

$$\frac{3}{5} \times \frac{2}{4} = \frac{6}{20} = \frac{3}{10}$$

因此，拿到至少一颗橙色弹珠的概率等于1减去拿到两颗蓝色弹珠的概率。

$$P(A) = 1 - P(\bar{A})$$

$$1-\frac{3}{10}=\frac{7}{10}$$

解答此题的另一个方法是使用直接法,即写出所有概率,然后把我们要求的概率相加。

随机选取两颗弹珠有四种可能的结果。其中三种结果会出现至少一颗橙色弹珠:

橙色,蓝色:$\frac{2}{5}\times\frac{3}{4}=\frac{6}{20}$

蓝色,橙色:$\frac{3}{5}\times\frac{2}{4}=\frac{6}{20}$ $\Biggr\}\frac{14}{20}\Rightarrow\frac{7}{10}$

橙色,橙色:$\frac{2}{5}\times\frac{1}{4}=\frac{2}{20}$

蓝色,蓝色:$\frac{3}{5}\times\frac{2}{4}=\frac{6}{20}$

注意上述概率的总和等于1(即,$\frac{6}{20}+\frac{6}{20}+\frac{2}{20}+\frac{6}{20}=\frac{20}{20}=1$),因此,除了上述四种结果以外不会有其他可能性。

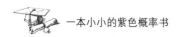

第4题 考试时间（一）

答案： E

解析： 题中使用了"或"这个词（即"通过第一门课程或第二门课程考试"），提示概率应该相加。计算两个非互斥事件 A 或 B 的概率，就是把第一个事件的概率与第二个事件的概率相加，然后减去两个事件之间的重复部分。这个被称为概率中的通用加法法则。见概率公式 #5（第 006 页）。

$$P(A+B) = P(A) + P(B) - P(AB)$$

$$\frac{9}{12} + \frac{8}{12} - \frac{6}{12} = \frac{11}{12}$$

在此，通过第一门课程考试的概率与通过第二门课程考试的概率相加，然后减去同时通过两门课程考试的概率。同时通过两门课

答案和解析

程考试的概率计算如下：$\frac{3}{4} \times \frac{2}{3} = \frac{6}{12} = \frac{1}{2}$。如果我们不减去这个部分，那么同时通过两门课程考试的概率会被计入两次。

> **注 意**
>
> 验证这个结果有一个方法，即明白通过任一门考试是指除两门课程考试都没有及格的其他任何情况。两门课程考试都不及格的概率是 $\frac{1}{4} \times \frac{1}{3} = \frac{1}{12}$。因此，通过任意一门课程考试（或两门考试）的概率是 $\frac{11}{12}$。

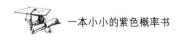

第 5 题　第六感

答案： D

解析： "或"的使用，说明需要把概率相加（即"第一次掷骰子或第二次掷骰子"）。接下来就是，是否需要减除重复。即，"这两个事件互斥吗？"根据题目我们知道，它们非互斥，所以重复部分必须减除。

让我们用图表展示掷两次骰子的所有可能性。

			第二次掷骰子			
	1	2	3	4	5	6
1	1, 1	1, 2	1, 3	1, 4	1, 5	1, 6
2	2, 1	2, 2	2, 3	2, 4	2, 5	2, 6
3	3, 1	3, 2	3, 3	3, 4	3, 5	3, 6
4	4, 1	4, 2	4, 3	4, 4	4, 5	4, 6
5	5, 1	5, 2	5, 3	5, 4	5, 5	5, 6
6	6, 1	6, 2	6, 3	6, 4	6, 5	6, 6

（第一次掷骰子为行标题）

答案和解析

如上图所示，掷一颗骰子两次（或同时掷两颗骰子）必然会有36种可能的结果。如表格中的数字所示，能得到6的结果有11种：（6，1）（6，2）（6，3）（6，4）（6，5）（6，6）和（1，6）（2，6）（3，6）（4，6）（5，6）。解答这道题运用的公式是概率的通用加法法则：

$$P(A+B) = P(A) + P(B) - P(AB)$$

$$\frac{1}{6} + \frac{1}{6} - \frac{1}{36}$$

$$= \frac{6}{36} + \frac{6}{36} - \frac{1}{36}$$

$$= \frac{12}{36} - \frac{1}{36} = \frac{11}{36}$$

> **注意**
>
> 不能直接用 $\frac{1}{6} + \frac{1}{6} = \frac{12}{36} / \frac{1}{3}$，而选择E。如果这样，就没有考虑并合理减除两次掷出6造成的重复。

第一次掷骰子得到6的概率是多少？

答案：（6，1）（6，2）（6，3）（6，4）（6，5）（6，6）。

事件A的概率是$\frac{6}{36}$或$\frac{1}{6}$。

第二次掷骰子得到6的概率是多少?

答案:(1,6)(2,6)(3,6)(4,6)(5,6)(6,6)。

事件B的概率是$\frac{6}{36}$或$\frac{1}{6}$。

注意在事件A和事件B中都包含了"双6",这个重复必须减除。需要说明的是,第一次掷出6或第二次掷出6的概率确实包括两次都掷出6的概率的情况(即双6),但这个结果只能计算一次。

这道题的另一种问法是:"掷两个正常的六面骰子,至少得到一个6的概率是多少?"

解答这道题的方法是使用概率的互补法则。"至少得到一个6的概率"等于"1减去没有掷出6的概率"。

没有掷出6的概率是:

$$\frac{5}{6} \times \frac{5}{6} = \frac{25}{36}$$

掷出至少一个6的概率是:

$$1 - \frac{25}{36} = \frac{11}{36}$$

> **注 意**
>
> 如果题目问:"掷一颗骰子两次,恰好得到一个6的概率是多少?"则选项B为正确答案。也许得出正确答案最简单的方法是写出所有的可能性。恰好得到一个6有10种情况:(6,1)(6,2)(6,3)(6,4)(6,5)(1,6)(2,6)(3,6)(4,6)和(5,6)。或者,我们也可以从之前的计算结果(即 $\frac{11}{36}$)中直接减去 $\frac{1}{36}$,即减去掷出双6的概率。注意,在下面的计算中,减去的第一个 $\frac{1}{36}$,代表重复掷出双6的概率,而减去的第二个 $\frac{1}{36}$,则代表掷出双6的实际概率。
>
> $$\frac{1}{6}+\frac{1}{6}-\frac{1}{36}-\frac{1}{36}=\frac{10}{36}=\frac{5}{18}$$

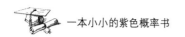

第6题 考试时间（二）

答案： E

解析： 这道题涉及三个重复概率，根据"快捷"算法的证明，最好是用概率的互补法则解答此题。

Ⅰ. 快捷算法

使用互补法则，我们要算出三门考试都不及格的概率。然后，用1减去三门考试都不及格的概率，从而得出通过任一或所有考试的概率。

未通过第一门课程考试的概率为：

$$P(\overline{A}) = 1 - P(A) \quad 1 - \frac{3}{4} = \frac{1}{4}$$

未通过第二门课程考试的概率：

$$P(\overline{B}) = 1 - P(B) \quad 1 - \frac{2}{3} = \frac{1}{3}$$

未通过第三门课程考试的概率

$$P(\overline{C}) = 1 - P(C) \quad 1 - \frac{1}{2} = \frac{1}{2}$$

三门课程考试全都未通过的概率为:

$$P(\overline{A}\,\overline{B}\,\overline{C}) \quad \frac{1}{4} \times \frac{1}{3} \times \frac{1}{2} = \frac{1}{24}$$

至少通过一门课程考试的概率为:

$$P(A) = 1 - P(\overline{A}) \quad 1 - \frac{1}{24} = \frac{23}{24}$$

Ⅱ. 直接算法

使用直接算法，分别计算出只通过一门课程考试的概率、通过两门课程考试的概率和通过全部考试的概率。然后，把七种结果相加。

通过第一门课程但未通过第二和第三门课程考试的概率：

$$P(A) \times P(\overline{B}) \times P(\overline{C}) \quad \frac{3}{4} \times \frac{1}{3} \times \frac{1}{2} = \frac{3}{24}$$

通过第二门课程但未通过第一和第三门课程考试的概率：

$$P(\overline{A}) \times P(B) \times P(\overline{C}) \quad \frac{1}{4} \times \frac{2}{3} \times \frac{1}{2} = \frac{2}{24}$$

通过第三门课程但未通过第一和第二门课程考试的概率：

$$P(\overline{A}) \times P(\overline{B}) \times P(C) \quad \frac{1}{4} \times \frac{1}{3} \times \frac{1}{2} = \frac{1}{24}$$

通过第一、第二门课程但未通过第三门课程考试的概率：

$$P(A) \times P(B) \times P(\overline{C}) \quad \frac{3}{4} \times \frac{2}{3} \times \frac{1}{2} = \frac{6}{24}$$

通过第一、第三门课程但未通过第二门课程考试的概率：

$$P(A) \times P(\overline{B}) \times P(C) \quad \frac{3}{4} \times \frac{1}{3} \times \frac{1}{2} = \frac{3}{24}$$

通过第二、第三门课程但未通过第一门课程考试的概率：

$$P(\overline{A}) \times P(B) \times P(C) \quad \frac{1}{4} \times \frac{2}{3} \times \frac{1}{2} = \frac{2}{24}$$

三门课程考试全部通过的概率：

$$P(A) \times P(B) \times P(C) \quad \frac{3}{4} \times \frac{2}{3} \times \frac{1}{2} = \frac{6}{24}$$

三门课程考试全都未通过的概率：

$$P(\overline{A}) \times P(\overline{B}) \times P(\overline{C}) \quad \frac{1}{4} \times \frac{1}{3} \times \frac{1}{2} = \frac{1}{24}$$

以上是一名学生参加考试的所有可能结果。把上述八项概率的前七项相加就能得出使用直接算法的正确答案。

证明：$\frac{3}{24} + \frac{2}{24} + \frac{1}{24} + \frac{6}{24} + \frac{3}{24} + \frac{2}{24} + \frac{6}{24} = \frac{23}{24}$

注意上述八种结果的和为1，因为1是所有概率的总和。

证明：$\frac{3}{24} + \frac{2}{24} + \frac{1}{24} + \frac{6}{24} + \frac{3}{24} + \frac{2}{24} + \frac{6}{24} + \frac{1}{24} = \frac{24}{24} = 1$

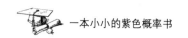

第7题 招聘

答案： A

解析： 这道题常被误认为排列题，但其实它不属于概率、排列、组合的范畴。

解这道题只需要把单个概率相乘。即把7（即销售经理候选人）、4（即，运务员候选人）和10（即，接待人员候选人）相乘，得到280种可能。

$$7 \times 4 \times 10 = 280$$

注意

这道题是关于一系列独立的选择。它运用了"乘数法则"，属于枚举法则范畴。排列公式不适用而且不能用于解答这类题目。这道题是关于我们有多少种选择，而不是有多少种排列的可能性，后者是处理排列题时的情况。

第8题 击剑

答案： C

解析： 这是一道排列题，不是组合题。在排列题中，顺序有影响。国家 A 赢得比赛、国家 B 第二和国家 B 赢得比赛、国家 A 第二的结果是不相同的。

$$P_n^r = \frac{n!}{(n-r)!}$$

$$P_4^2 = \frac{4!}{(4-2)!} = \frac{4!}{2!} = \frac{4 \times 3 \times 2 \times 1}{2 \times 1} = 12$$

> **注意**
>
> 思考下列补充问题。一名老师的特别补习班里有 4 名学生。

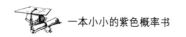一本小小的紫色概率书

期末,这名老师将颁发4个奖项——数学、英语、历史和创造性写作奖。假设一名学生可以获得多个奖项,那么老师有多少种颁奖方式?

$$n^r=4^4=4\times4\times4\times4=256$$

这名老师有4种方式颁发数学奖,4种方式颁发英语奖,4种方式颁发历史奖,4种方式颁发创造性写作奖。参见概率公式#10(第014页)。

第9题 座次（一）

答案：E

解析：这是一道排列题，适合用快捷公式——$n!$。因为我们需要用到集合 n 的所有成员，而不是集合 n 的子集。

这道题实际上是问："有多少种方法在 6 个座位上安排 6 个人？"这 6 个人中有几名男孩和几名女孩与本题没有关系。如果对学生的座次安排没有限定，那么他们有 720 种可能的坐法。

$$6! = 6 \times 5 \times 4 \times 3 \times 2 \times 1 = 720$$

有 6 种方法安排第一名学生的座位，5 种方法安排第二名学生的座位，4 种方法安排第三名学生的座位，3 种方法安排第四名学生的座位，2 种方法安排第五名学生的座位，只有 1 种方法安排第六名也就是最后一名学生的座位。

第10题 座次（二）

答案：C

解析：这实际上是一道联合排列题，算出两个单独的概率后再把结果相加。

女孩和男孩的补考座次安排有两种可能。第一种场景：第一、三、五个座位坐男孩，第二、四、六个座位坐女孩。第二种场景：第一、三、五个座位坐女孩，第二、四、六个座位坐男孩。（B 男孩，G 女孩）

第一种场景：

$$\begin{array}{cccccc} B & G & B & G & B & G \end{array}$$

$$\frac{3}{B_1} \times \frac{3}{G_1} \times \frac{2}{B_2} \times \frac{2}{G_2} \times \frac{1}{B_3} \times \frac{1}{G_3}$$

第二种场景：

$$G \quad B \quad G \quad B \quad G \quad B$$

$$\frac{3}{G_1} \times \frac{3}{B_1} \times \frac{2}{G_2} \times \frac{2}{B_2} \times \frac{1}{G_3} \times \frac{1}{B_3}$$

对于第一种场景，每个座位分别有几种坐法（从左到右）？

答：第一个座位可能坐3位男孩的任一位，第二个座位可能坐3位女孩的任一位，第三个座位可能坐剩下的2位男孩的任一位，第四个座位可能坐剩下的2位女孩的任一位，第五个座位坐最后一位男孩，第六个座位坐最后一位女孩。

对于场景2，每个座位分别有几种坐法（从左到右）？

答：第一个座位可能坐3位女孩的任一位，第二个座位可能坐3位男孩的任一位，第三个座位可能坐剩下的2位女孩的任一位，第四个座位可能坐剩下的2位男孩的任一位，第五个座位坐最后一位女孩，第六个座位坐最后一位男孩。

因此：

$$B\ G\ B\ G\ B\ G \quad G\ B\ G\ B\ G\ B$$

$$(3 \times 3 \times 2 \times 2 \times 1 \times 1) + (3 \times 3 \times 2 \times 2 \times 1 \times 1)$$

$$= 36 + 36 = 72$$

简言之，可以按如下方法算出答案：

$$(3!\times 3!)+(3!\times 3!)$$

$$=2\times(3!\times 3!)$$

$$=2\times\left[(3\times 2\times 1)\times(3\times 2\times 1)\right]=72$$

$$=2\times(6\times 6)=72$$

$$=2\times 36=72$$

> **注 意**
>
> 这种类型的排列题还衍生出一种常见变体：
>
> 3位男孩和3位女孩准备安排座次参加补考。女孩必须坐在第一、二、三个座位，男孩必须坐在第四、五、六个座位。那么，6名学生的座位安排有多少种可能？
>
> $$B \quad G \quad B \quad G \quad B \quad G$$
>
> $$\frac{3}{B_1}\times\frac{3}{G_1}\times\frac{2}{B_2}\times\frac{2}{G_2}\times\frac{1}{B_3}\times\frac{1}{G_3}$$
>
> **答案**：有 $3!\times 3!=6\times 6=36$ 种可能性。

答案和解析

第11题 Banana

答案：D

解析：这道题突出了"重复字母"（或"重复数字"）的处理。含重复数字或字母的排列的计算公式是 $\dfrac{n!}{x!y!z!}$，x、y、z 指存在重复字母或数字的不同字母或数字。

$$\dfrac{n!}{x!y!} = \dfrac{6!}{3! \times 2!} = \dfrac{6 \times 5 \times 4 \times 3 \times 2 \times 1}{3 \times 2 \times 1 \times 2 \times 1} = 60$$

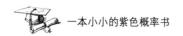

第 12 题　餐桌

答案：C

解析：这道题涉及棘手的"空座位"问题的处理。

$$\frac{5!}{2!} = \frac{5\times 4\times 3\times 2\times 1}{2\times 1} = 60$$

2! 代表两个空座位。

注　意

　　这道题的答案与第 11 题 *Banana* 的解法相似。排列理论中，"空座位"与"相同数字"（或"相同字母"）类似。想象这两把空椅子代表两位相同的人。另外，圆桌的几何形状不应成为干扰因素。解答这道题与解答并排的五把椅子问题是一样的。要知道，餐桌不过是一排椅子首尾相接，就圆桌而言则形同一个圆圈。

第13题　歌手

答案：C

解析：解联合组合题要把两个单独的组合结果相乘。

首先，把组合题分解成两项运算。一是"老歌"，二是"新歌"。

老歌：

$$C_n^r = \frac{n!}{r!(n-r)!}$$

$$C_6^4 = \frac{6!}{4!(6-4)!} = \frac{6!}{4!(2!)} = \frac{6\times5\times4\times3\times2\times1}{4\times3\times2\times1\times(2\times1)} = 15$$

故，15代表歌手从6首老歌中选择4首歌演唱的可能选法。

新歌：

$$C_n^r = \frac{n!}{r!(n-r)!}$$

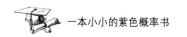

$$C_5^2 = \frac{5!}{2!(5-2)!} = \frac{5!}{2!(3!)} = \frac{5 \times 4 \times 3 \times 2 \times 1}{2 \times 1 \times (3 \times 2 \times 1)} = 10$$

故，10 代表歌手从 5 首新歌中选择 2 首歌演唱的可能选法。

因此，联合组合等于 $15 \times 10 = 150$。

简言之，这道联合组合题的结果为：

$$C_n^r \times C_n^r = \frac{n!}{r!(n-r)!} \times \frac{n!}{r!(n-r)!}$$

$$C_6^4 \times C_5^2 = \frac{6!}{4!(6-4)!} \times \frac{5!}{2!(5-2)!}$$

$$C_6^4 \times C_5^2 = \frac{6!}{4!(2)!} \times \frac{5!}{2!(3)!}$$

$$C_6^4 \times C_5^2 = \frac{6 \times 5 \times 4!}{4!(2!)} \times \frac{5 \times 4 \times 3!}{2!(3!)} = 15 \times 10 = 150$$

答案和解析

第14题 重逢

答案: D

解析: 表面上看,这道题好像非常复杂,但它的解法其实很简单。题目其实是问:"一个11人的群体,顺序不影响,两两组合可以组多少对?"或者更明确一些:"从11个人中选择2个人,顺序不影响,有多少种选法?"

$$C_n^r = \frac{n!}{r!(n-r)!}$$

$$C_{11}^2 = \frac{11!}{2!(11-2)!} = \frac{11 \times 10 \times 9!}{2(9!)} = 55$$

> **注 意**
>
> 这道题如果这样问:"为了举行沙滩排球比赛,11个人可以组多少个两人小组?"答案也是一样的。组队是典型的组合范例,因为顺序不影响。小组成员无论先选还是后选都没有关系。

第15题 结果

答案：A

解析：这道题测试了最基础的排列和组合理论。如果对理论已经深刻理解，就不需进行任何计算。

表述Ⅰ：

正确。$P_5^3 > P_5^2$

$P_5^3 = 60$，而 $P_5^2 = 20$。顺序对排列有影响，而且排列项目越多，可能性越多。

表述Ⅱ：

错误。$C_5^3 > C_5^2$

$C_5^3 = 10$，$C_5^2 = 10$。结果竟然相等！"组合中的互补"使两种结果相等。本题中 $3+2=5$。注意，这种现象只在组合中发生，不会在排列中发生。

表述Ⅲ：

错误。$C_5^2 > P_5^2$

因为，$C_5^2 = 10$，而 $P_5^2 = 20$。顺序影响排列，这会使排列产生相较组合而言更多的可能性。反过来说也就是，所有条件相同的情况下，因为顺序不影响组合，这使组合的结果比排列的结果少。

哈佛大学毕业生撰写的逻辑入门书。
美国常春藤名校入学考试必备读物!
一本能让你思维"如刀锋般犀利"的逻辑入门书!

书名:《一本小小的蓝色逻辑书》
作者:(加)布兰登·罗伊尔
出版社:九州出版社
定价:39.80元

★ 一本哈佛大学毕业生写的逻辑入门书。

★ 美国常春藤名校入学考试绝密利器,先后荣获"美国总统图书奖"、"国际图书奖"、"年度教育图书奖"!

★ 美国著名考试培训机构Kaplan教学主管20年教学经验总结。

★ 一套完整而实用的逻辑推理概念,囊括50个逻辑推理法则,帮助你用逻辑思维解决现实问题。一套针美国大学入学考试的阅读理解和逻辑分析题解题思路,并辅以40多道实战题目供你反复练习。

Contents

Introduction, 1

Overview, 2

Probability, 15

Enumerations, 17

Permutations, 18

Combinations, 19

Answers and Explanationa, 20

About the Author, 38

Introduction

This guidebook is compiled with one primary goal in mind: To help readers master basic probability in the shortest possible time frame.

Achieving this goal requires two steps. The first step is to develop a superior conceptual framework for understanding probability. The overview section at the beginning of this material contains two strategic flowcharts—one for probability and one for permutations and combinations—the two main topics under review. Included is a summary of all 14 essential formulas, each accompanied by a relevant example. The second step is to practice on a small number of the most classic, thematic problems that represent the broader types of probability problems that may be found elsewhere.

Mastering the thinking skills to succeed in basic probability requires an ability to match a given problem with the correct formula. In order to properly categorize a problem, it is necessary to understand the often confusing terminology that surrounds probability and its related topics. For example, what does it mean to say that two events are independent or that they are not mutually exclusive? Once we are confident that we can sort problems into their common types, we can then judge whether a particular problem is a variation based on a classic problem type and/or whether it contains common traps that are otherwise easy to overlook. Special notes (marked NOTE ᛋ) are imbedded throughout this material. Here the reader will find additional information, examples, or follow-up problems that will no doubt prove useful.

This guidebook provides a complement to *The Little Green Math Book*, which addresses basic but tricky math involving arithmetic and algebra. Probability is a subject that warrants special attention and this book serves as a stand-alone workshop.

Let's get started.

Overview

This material covers three main topics: probability, permutations, and combinations. What is the difference between probability and permutations and combinations? Probabilities are expressed as decimals or fractions between 0 and 1 (where 1 is the probability of certainty and 0 is the probability of impossibility) or, alternatively, as percentages between 0% and 100% (where 100% is the probability of certainty and 0% is the probability of impossibility). Permutations and combinations, on the other hand, result in outcomes greater than or equal to 1. Frequently they result in quite large outcomes such as 10, 36, 720, etc.

Probability:

In terms of probability, a quick rule of thumb is to determine first whether we are dealing with an "and" or "or" situation. "And" means multiply and "or" means addition. For example, if a problem states "what is the probability of x and y," we multiply individual probabilities together. If a problem states "what is the probability of x or y," we add individual probabilities together.

Moreover, if a probability problem requires us to multiply, we must ask one further question: "Are the events independent or are the events dependent?" *Independent* means that two events have no influence on one another and we simply multiply individual probabilities together to arrive at a final answer. *Not independent (dependent)* means that the occurrence of one event has an influence on the occurrence of another event and this influence must be taken into account.

Likewise, if a problem requires us to add probabilities, we must ask one further question: "Are the events mutually exclusive or non-mutually exclusive?" *Mutually exclusive* means that two events cannot occur at the

same time and there is no "overlap" present. If two events have no overlap, we simply add probabilities. *Not mutually exclusive* means that two events can occur at the same time and that overlap is present. If two events do contain overlap, this overlap must not be double counted.

Permutations and Combinations:

With respect to permutations and combinations, permutations are ordered groups while combinations are unordered groups. Order matters in permutations; order does not matter in combinations. For example, AB and BA are considered different outcomes in permutations, but they are considered a single outcome in combinations. In real-life, examples of permutations include telephone numbers, license plates, electronic codes, and passwords. Examples of combinations include selection of members for a team or lottery tickets. In the case of lottery tickets, for instance, the order of numbers does not matter; we just need to get all the numbers, usually six of them.

As a practitioner's rule, the words "arrangements" or "possibil- ities" imply permutations; the words "select" or "choose" imply combinations.

Factorials:

Factorial means that we engage multiplication such that:
    Example: $4! = 4 \times 3 \times 2 \times 1$
    Example: $7! = 7 \times 6 \times 5 \times 4 \times 3 \times 2 \times 1$
Zero factorial equals one and one factorial also equals one:
    Example: $0! = 1$
    Example: $1! = 1$

Coins, Dice, Marbles, and Cards:

Problems in this guidebook include reference to coins, dice, marbles, and cards. For clarification purposes: The two sides of a coin are heads and tails. A die has six sides numbered from 1 to 6, with each having an equal likelihood of appearing subsequent to being tossed. The word "die" is singular; "dice" is plural. Marbles are assumed to be of a single, solid color. A deck of cards contains 52 cards divided equally into four suits—clubs, diamonds, hearts, and spades—where each suit contains 13 cards including ace, king, queen, jack, 10, 9, 8, 7, 6, 5, 4, 3, and 2.

OVERVIEW

Exhibit A   Probability Flowchart

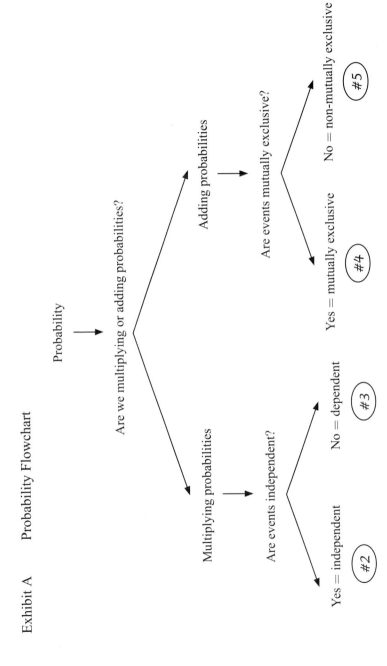

The numbers above denote the applicable probability formula.

THE LITTLE PURPLE PROBABILITY BOOK

Exhibit B   Permutations and Combinations Flowchart

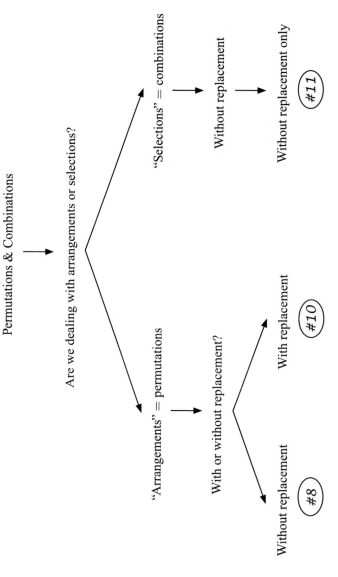

The numbers above denote the applicable permutation or combination formula.

Basic Probability Formulas

Here at a glance are the basic probability, permutation, and combination formulas used in this material.

Universal Formula

#1 $\quad \text{Probability} = \dfrac{\text{Selected Events(s)}}{\text{Total Number of Possibilities}}$

Example: You buy 3 raffle tickets and there are 10,000 tickets sold. What is the probability of winning the single prize?

$$\text{Probability} = \dfrac{3}{10,000}$$

Special Multiplication Rule

#2 $\quad P(A \text{ and } B) = P(A) \times P(B)$

[Where the probability of A and B equals the probability of A times the probability of B]

If events are independent (that is, one event has no influence on the other), we simply multiply them together.

Example: What is the probability of tossing a coin twice and obtaining heads on both the first and second toss?

$$\dfrac{1}{2} \times \dfrac{1}{2} = \dfrac{1}{4}$$

General Multiplication Rule

#3 $\quad P(A \text{ and } B) = P(A) \times P(B/A)$

[Where the probability of A and B equals the probability of A times the probability of B, given that A has already occurred]

If events are not independent (that is, one event has an

influence on the other), we must adjust the second event based on its influence from the first event.

Example: A bag contains six marbles, three blue and three green. What is the probability of blindly reaching into the bag and pulling out two green marbles?

$$\frac{3}{6} \times \frac{2}{5} = \frac{6}{30} = \frac{1}{5}$$

Special Addition Rule

#4   P(A or B) = P(A) + P(B)

[Where the probability of A or B equals the probability of A added to the probability of B]

If events are mutually exclusive (that is, there is no overlap), then we just add the probability of the events together.

Example: The probability that Sam will go to high school at Montcalm Academy is 50 percent, while the probability that he will go to Crescent Heights High School is 25 percent. What is the probability that he will choose to go to high school at either Montcalm Academy or Crescent Heights High School?

$$50\% + 25\% = 75\%$$

General Addition Rule

#5   P(A or B) = P(A) + P(B) − P(A and B)

[Where the probability of A or B equals the probability of A added to the probability of B minus the probability of A and B]

If events are not mutually exclusive (that is, there is overlap), then we must subtract out the overlap subsequent to adding the events.

Example: The probability that tomorrow will be rainy is 30 percent. The probability that tomorrow will be windy is 20 percent. What is the probability that tomorrow's weather will be either rainy or windy?

$$30\% + 20\% - (30\% \times 20\%)$$
$$30\% + 20\% - 6\%$$
$$50\% - 6\% = 44\%$$

With regard to the General Addition Rule, the reason that we subtract out the overlap is because we do not want to count it twice. When two events overlap, both events contain that same overlap. Thus, it must be subtracted once in order not to "double" count it.

NOTE ∽ Let's quickly contrast what is commonly referred to as the inclusive "or" and the exclusive "or." The problem used in support of probability formula #5 is governed by an inclusive "or." It is reasonable to assume that tomorrow's weather can be both rainy and windy. The problem is effectively asking: "What is the probability that tomorrow's weather will be either rainy, or windy, or both rainy and windy?" The inclusive "or" occurs whenever there is overlap. The problem used in support of probability formula #4 effectively asks: "What is the probability that Sam will choose to go to high school at Montcalm Academy or Crescent Heights High School, but not both high schools?" The choice between going to high school at two different locations is a mutually exclusive one, and we treat that particular problem as involving an exclusive "or."

Complement Rule

#6 $P(A) = 1 - P(\text{not } A)$

[Where the probability of A equals one minus the probability of A not occurring]

The Complement Rule of probability describes the subtracting of probabilities rather than the adding or multiplying of probabilities. To calculate the probability of an event using this rule, we ask: "What is the probability of a given event not occurring?" Then we subtract this result from 1.

Example: What is the probability of rolling a pair of dice and not rolling double sixes?

The probability of rolling double sixes:

$$\frac{1}{6} \times \frac{1}{6} = \frac{1}{36}$$

The probability of not rolling double sixes:

$$1 - \frac{1}{36} = \frac{35}{36}$$

Rule of Enumeration

#7 If there are x ways of doing one thing, y ways of doing a second thing, and z ways of doing a third thing, then the number of ways doing all these things is x × y × z. This is known as the Rule of Enumeration.

NOTE ∽ Technically, the Rule of Enumeration does not fall under the umbrella of "probability" or "permutation" or "combination." For practical reasons, however, it is most often discussed along with probability.

Example: Fast-Feast Restaurant offers customers a set menu with a choice of one of each of the following: 2 different salads, 3 different soups, 5 different entrees, 3 different desserts, and coffee or tea. How many possibilities are there with respect to how a customer can take his or her meal?

$$2 \times 3 \times 5 \times 3 \times 2 = 180$$

OVERVIEW

Permutations

#8 Without replacement $P_n^r = \dfrac{n!}{(n-r)!}$

[Where n = total number of items and r = number of items we are taking or arranging]

Example: How many two-letter codes can be made from the letters A, B, C, and D if the same letter cannot be displayed more than once in any given code?

$$P_n^r = \dfrac{n!}{(n-r)!}$$

$$P_4^2 = \dfrac{4!}{(4-2)!} = \dfrac{4!}{2!} = \dfrac{4 \times 3 \times 2 \times 1}{2 \times 1} = 4 \times 3 = 12$$

#9 $P_n^n = n!$

[Shortcut formula when all items are taken together]

Example: How many ways can a person display (or arrange) four different books on a shelf?

$$P_n^r = \dfrac{n!}{(n-r)!}$$

$$P_4^4 = \dfrac{4!}{(4-4)!} = \dfrac{4!}{0!} = \dfrac{4}{1} = 4! = 4 \times 3 \times 2 \times 1 = 24$$

Also, shortcut formula: $n! = 4! = 24$

#10 With replacement n^r

Example: How many four-digit codes can be made from the numbers 1, 2, 3, and 4, if the same numbers can be displayed more than once in any given code?

$$n^r \quad 4^4 = 256$$

NOTE ∽ Permutation with replacement (that is, n^r) technically falls under the Rule of Enumeration. It is included here for ease of presentation. For a problem to be considered a permutation, the permutation formula must be applicable.

Combinations

(#11) $C_n^r = \dfrac{n!}{r!(n-r)!}$

[Where n = total number of items taken and r = the number of items we are choosing or selecting]

Example: How many ways can a person choose three of four colors for the purpose of painting the inside of a house?

$$C_n^r = \dfrac{n!}{r!(n-r)!}$$

$$C_4^3 = \dfrac{4!}{3!(4-3)!} = \dfrac{4!}{3!(1)!} = \dfrac{4 \times 3 \times 2 \times 1}{3 \times 2 \times 1 \times 1} = 4$$

Additional formulas

Joint Permutations

$$\boxed{\#12} \quad P_n^r \times P_n^r = \frac{n!}{(n-r)!} \times \frac{n!}{(n-r)!}$$

Example: A tourist plans to visit three of five Western European cities and then proceed to visit two of four Eastern European cities. At the planning stage, how many itineraries are possible?

$$P_5^3 \times P_4^2$$

$$\frac{5!}{(5-3)!} \times \frac{4!}{(4-2)!}$$

$$\frac{5!}{2!} \times \frac{4!}{2!}$$

$$\frac{5 \times 4 \times 3 \times 2 \times 1}{2 \times 1} \times \frac{4 \times 3 \times 2 \times 1}{2 \times 1}$$

$$60 \times 12 = 720$$

Multiplying outcomes, rather than adding them, is consistent with the treatment afforded by the Rule of Enumeration.

Joint Combinations

$$\boxed{\#13} \quad C_n^r \times C_n^r = \frac{n!}{r!(n-r)!} \times \frac{n!}{r!(n-r)!}$$

Example: A special marketing task force is to be chosen from five professional golfers and five professional tennis players. If the final task force chosen is to consist of three golfers and three tennis players, then how many different task forces are possible?

$$C_5^3 \times C_5^3$$

$$\frac{5!}{3!(5-3)!} \times \frac{5!}{3!(5-3)!}$$

$$\frac{5\times4\times3\times2\times1}{3\times2\times1\times2\times1} \times \frac{5\times4\times3\times2\times1}{3\times2\times1\times2\times1}$$

$$10\times10=100$$

Repeated Letters or Numbers (Permutations)

#14 $\dfrac{n!}{x!y!z!}$

[where x, y and z are different but identical letters or numbers]

Example: How many four-numeral codes can be created using the four numbers 0, 0, 1 and 2?

$$\frac{4!}{2!}=\frac{4\times3\times2\times1}{2\times1}=12$$

Note that 2! denotes the two zeros which represent repeated numbers.

Probability

1. Four Aces

Which of the following represents the probability of selecting four cards at random from a deck of cards and getting four aces? (The cards are to be selected one after the other without replacing any of the cards.)

A) $\dfrac{1}{52} \times \dfrac{1}{52} \times \dfrac{1}{52} \times \dfrac{1}{52}$ B) $\dfrac{1}{52} \times \dfrac{1}{51} \times \dfrac{1}{50} \times \dfrac{1}{49}$

C) $\dfrac{4}{52} \times \dfrac{3}{51} \times \dfrac{2}{50} \times \dfrac{1}{49}$ D) $\dfrac{4}{52} \times \dfrac{3}{52} \times \dfrac{2}{52} \times \dfrac{1}{52}$

E) $\dfrac{4}{52} \times \dfrac{4}{52} \times \dfrac{4}{52} \times \dfrac{4}{52}$

2. Orange & Blue

There are 5 marbles in a bag—2 are orange and 3 are blue. If two marbles are pulled from the bag, what is the probability that the first will be orange and the second will be blue?

A) $\dfrac{6}{25}$ B) $\dfrac{3}{10}$ C) $\dfrac{2}{5}$ D) $\dfrac{3}{5}$ E) $\dfrac{7}{10}$

3. Orange & Blue Again

There are 5 marbles in a bag—2 are orange and 3 are blue. If two marbles are pulled from the bag, what is the probability that at least one will be orange?

A) $\dfrac{6}{25}$ B) $\dfrac{3}{10}$ C) $\dfrac{2}{5}$ D) $\dfrac{3}{5}$ E) $\dfrac{7}{10}$

4. Exam Time

A student is to take her final exams in two subjects. The probability that she will pass the first subject is $\dfrac{3}{4}$ and the probability that she will pass the second subject is $\frac{2}{3}$.

What is the probability that she will pass one exam or the other exam?

A) $\frac{5}{12}$ B) $\frac{1}{2}$ C) $\frac{7}{12}$ D) $\frac{5}{7}$ E) $\frac{11}{12}$

5. Sixth Sense

What is the probability of rolling a six on either the first or second toss of a dice?

A) $\frac{1}{36}$ B) $\frac{5}{18}$ C) $\frac{1}{6}$ D) $\frac{11}{36}$ E) $\frac{1}{3}$

6. Exam Time Encore

A student is to take her final exams in three subjects. The probability that she will pass the first subject is $\frac{3}{4}$, the probability that she will pass the second subject is $\frac{2}{3}$, and the probability that she will pass the third subject is $\frac{1}{2}$. What is the probability that she will pass at least one of these three exams? (Or: What is the probability that she will pass either the first exam or the second exam or the third exam?)

A) $\frac{1}{4}$ B) $\frac{11}{24}$ C) $\frac{17}{24}$ D) $\frac{3}{4}$ E) $\frac{23}{24}$

Enumerations

7. Hiring

A company seeks to hire a sales manager, a shipping clerk, and a receptionist. The company has narrowed its candidate search and plans to interview all remaining candidates including 7 persons for the position of sales manager, 4 persons for the position of shipping clerk, and 10 persons for the position of receptionist. How many different hirings of these three people are possible?

A) $7 \times 4 \times 10$ B) $7 + 4 + 10$ C) $21 \times 20 \times 19$

D) $7! + 4! + 10!$ E) $7! \times 4! \times 10!$

THE LITTLE PURPLE PROBABILITY BOOK

Permutations

8. Fencing

Four contestants representing four different countries advance to the finals of a fencing championship. Assuming all competitors have an equal chance of winning, how many possibilities are there with respect to how a first-place and second-place medal can be awarded?

A) 6   B) 7   C) 12   D) 16   E) 24

9. Row

Six students are to sit in a row side by side for a makeup exam. How many ways could they arrange themselves?

A) 12   B) 36   C) 72   D) 240   E) 720

10. Alternating

Six students—3 boys and 3 girls—are to sit in a row side by side for a makeup exam. How many ways could they arrange themselves given that no two boys and no two girls can sit next to one another?

A) 12   B) 36   C) 72   D) 240   E) 720

11. Banana

Which of the following leads to the correct mathematical solution for the number of ways that the letters of the word BANANA could be arranged to create a six-letter code?

A) $6!$ B) $6! - (3! \times 2!)$ C) $6! - (3! + 2!)$

D) $\dfrac{6!}{3! \times 2!}$ E) $\dfrac{6!}{3! + 2!}$

12. Table

How many ways could three people sit at a table with five seats in which two of the five seats will remain empty?

A) 8   B) 12   C) 60   D) 118   E) 120

Combinations

13. Singer

For an upcoming charity event, a male vocalist has agreed to sing 4 out of 6 "old songs" and 2 out of 5 "new songs." How many ways can the singer make his selection?

A) 25   B) 50   C) 150   D) 480   E) 600

14. Reunion

If 11 people meet at a reunion and each person shakes hands exactly once with each of the others, what is the total number of handshakes?

A) $11\times10\times9\times8\times7\times6\times5\times4\times3\times2\times1$

B) $10\times9\times8\times7\times6\times5\times4\times3\times2\times1$

C) 11×10

D) 55

E) 45

15. Outcomes

Given that $P_n^r = \dfrac{n!}{(n-r)!}$ and $C_n^r = \dfrac{n!}{r!(n-r)!}$, where n is the total number of items and r is the number of items taken or chosen, which of the following statements are true in terms of the number of outcomes generated?

I . $P_5^3 > P_5^2$ II . $C_5^3 > C_5^2$ III . $C_5^2 > P_5^2$

A) 只有 I   B) 只有 I 和 II   C) 只有 I 和 III

D) 只有 II 和 III   E) I 和 II 和 III

Answers and Explanations

1. Four Aces

Choice C

Overview: This problem presents a classic application of the General Multiplication Rule of probability. Selecting a card "without replacement" affects or influences the probability of the next card being chosen because the first card is missing from the deck. Selecting a card "with replacement" does not affect or influence the probability of the next card being chosen because replacing a given card restores the deck to its prior state.

On the first pick, there is a $\frac{4}{52}$ chance of selecting an ace. On our second pick, there is a $\frac{3}{51}$ chance of selecting an ace because there is one fewer ace to choose from and one fewer card in the deck. On the third pick, there is a $\frac{2}{50}$ chance of selecting an ace because there is one fewer ace to choose from and one fewer card in the deck. Finally, there is a $\frac{1}{49}$ chance of selecting an ace because there is only one ace to choose from with exactly 49 cards left in the deck.

For the record, answer choice E would have been the correct answer if we had selected our four aces with replacement. Such would be the case if the problem had asked: "What is the probability of selected four aces in a row if we replace each card before selecting the next one?" Choice A would be the correct answer if we were to choose any single card with replacement. For example: "What is the probability of selecting the ace of spades four times in a row from a random deck of cards, if we replace that card in the deck after selecting it?" Choice B would have been the correct answer if we had to select these four aces in a specific order and had done so without replacing each card. Such would be the case if the problem had asked: "What is the probability of selecting the ace of spades, followed by the ace of hearts, followed by the ace of clubs, and followed by the ace of diamonds, if we do not replace each card after selecting it?"

NOTE ∽ There are two sets of terms that are key to understanding how to distinguish among basic probability problems. The first set of terms is "mutually exclusive" and "not mutually exclusive." The second set of terms is "independent" and "not independent (dependent)." Mutually exclusive means that two events or outcomes do not overlap with one another or cannot occur at the same time. Not mutually exclusive means that two events or outcomes do overlap with one another or can occur at the same time. Independent means that two events or outcomes do not influence one another and occur randomly relative to each other. Not independent (or dependent) means that two events or outcomes influence one another and that the occurrence of one event has an affect on the occurrence of another event.

Here are some simple real-life examples to illuminate these terms. Say we are putting on a business conference and inviting attendees as well as guest speakers. The assignment of VIP seating and non-VIP seating is a mutually exclusive outcome. Either a person is in the VIP seats or he or she is not. The same holds true for determining who is an in-state versus out-of-state attendee. An attendee is either in-state or out-of-state with no overlap possible. However, in classifying attendees by profession, we might have overlap between who is a manager and who is an engineer and who is a salesperson and who is an entrepreneur. Obviously, some attendees might fall into more than one category. These categories would, therefore, not be mutually exclusive.

Two tasks might be independent and have no influence on one another. For example, in preparing for the conference, it wouldn't make any difference whether we made name tags first and then made copies of the conference handouts or made copies of the conference handouts and then made name tags. These events represent separate tasks that have no bearing on one another. On the other hand, two tasks may not be independent of one another; they may, in fact, be dependent on one another. This is true of events that must occur in a certain sequence. In preparing for the conference, we must plan the conference first before inviting speakers to speak at the conference. Likewise, attendees must be registered for the conference, before they can be admitted to the conference and before they ever fill out conference evaluation forms, which are handed out at the end of the conference. In other words, the filling out of a conference evaluation form is dependent upon a person actually attending the conference, which, in turn, is dependent upon a person having first registered for the conference.

2. Orange & Blue

Choice B

Overview: This problem is solved using the General Multiplication Rule. The use of the word "and" in this problem is a signal that we need to multiply probabilities, not add probabilities. The probability of choosing the first orange marble will influence the probability of choosing the first blue marble because there will be one fewer marble to choose from (selection without replacement).

The probability of first choosing an orange marble is $\frac{2}{5}$. The probability of then choosing a blue marble is $\frac{3}{4}$.

ANSWERS AND EXPLANATIONS

$$\frac{2}{5} \times \frac{3}{4} = \frac{6}{20} = \frac{3}{10}$$

Note that the trap answer, choice A, involves forgetting to subtract one marble from the denominator of the second fraction:

$$\frac{2}{5} \times \frac{3}{5} = \frac{6}{25}$$

3. Orange & Blue Again

Choice E

Overview: This problem introduces the Complement Rule of probability. See probability formula #6 (page 14).

The simplest way to view this problem is in terms of what we don't want. At least one orange marble means that we want anything but two blue marbles.

Probability of getting two blue marbles:

$$\frac{3}{5} \times \frac{2}{4} = \frac{6}{20} = \frac{3}{10}$$

Therefore, the probability of getting at least one orange marble is the same as one minus the probability of getting two blue marbles.

$$P(A) = 1 - P(\text{not } A)$$

$$1 - \frac{3}{10} = \frac{7}{10}$$

Another way to solve this problem is using the direct method, which entails writing out all possibilities and adding up the results that we are looking for. There are four possible outcomes when we choose two marbles at random. Three of these outcomes yield at least one orange marble:

Orange, Blue: $\dfrac{2}{5} \times \dfrac{3}{4} = \dfrac{6}{20}$

Blue, Orange: $\dfrac{3}{5} \times \dfrac{2}{4} = \dfrac{6}{20}$

$\left. \begin{array}{c} \\ \\ \\ \end{array} \right\} \dfrac{14}{20} \Rightarrow \dfrac{7}{10}$

Orange, Orange: $\dfrac{2}{5} \times \dfrac{1}{4} = \dfrac{2}{20}$

Blue, Blue: $\dfrac{3}{5} \times \dfrac{2}{4} = \dfrac{6}{20}$

Note that the total of all of the above possibilities equals 1 (that is, $\dfrac{6}{20} + \dfrac{6}{20} + \dfrac{2}{20} + \dfrac{6}{20} = \dfrac{20}{20} = 1$) because there are no other possibilities other than the four outcomes presented here.

4. Exam Time

Choice E

Overview: The use of the word "or" in this problem (that is, "pass one exam or the other exam") is a signal that probabilities are to be added, not multiplied. The probability of two non-mutually exclusive events A or B is calculated by adding the probability of the first event to the second event and then subtracting out the overlap between the two events. This is referred to in probability as the General Addition Rule. See probability formula #5 (page 13).

$$P(A \text{ or } B) = P(A) + P(B) - P(A \text{ and } B)$$

$$\dfrac{9}{12} + \dfrac{8}{12} - \dfrac{6}{12} = \dfrac{11}{12}$$

Here, the probability of passing the first exam is added to the probability of passing the second exam, less the probability of passing both exams. The

ANSWERS AND EXPLANATIONS

probability of passing both exams is calculated as follows: $\frac{3}{4} \times \frac{2}{3} = \frac{6}{12} = \frac{1}{2}$. If we don't make this subtraction, we will double count the possibility that she will pass both exams.

NOTE ✒ One way to prove this result is to recognize that the probability of passing either exam (including passing both exams) is everything other than failing both exams. The probability of failing both exams is $\frac{1}{4} \times \frac{1}{3} = \frac{1}{12}$. Therefore, the probability of passing either the first exam or the second exam (or both exams) is $\frac{11}{12}$.

5. Sixth Sense

Choice D

Overview: The use of the word "or" signals the need to add probabilities (that is, "first or second toss of a dice"). The only other question to be asked is whether there is a need to subtract out overlap. That is, "Are the two events mutually exclusive?" As it turns out, they are not mutually exclusive and this overlap must be subtracted out.

Let's use a chart to visualize all the possibilities that occur when a dice is rolled twice.

		Second Roll					
		1	2	3	4	5	6
	1	1,1	1,2	1,3	1,4	1,5	1,6
	2	2,1	2,2	2,3	2,4	2,5	2,6
First Roll	3	3,1	3,2	3,3	3,4	3.5	3,6
	4	4,1	4,2	4,3	4,4	4,5	4,6
	5	5,1	5,2	5,3	5,4	5,5	5,6
	6	6,1	6,2	6,3	6,4	6,5	6,6

As seen in the chart above, there are of course thirty-six possible outcomes when we roll a single dice twice (or, equally, if we toss a pair of dice). As indicated by the bolded numbers in the chart, we have 11 outcomes with respect to how we can get exactly one six: (6,1), (6,2), (6,3), (6,4), (6,5), (6,6) and (1,6), (2,6), (3,6), (4,6), (5,6).

The applicable formula for solving this particular problem is the General Addition Rule of probability:

$$P(A \text{ or } B) = P(A) + P(B) - P(A \text{ and } B)$$

$$\frac{1}{6} + \frac{1}{6} - \frac{1}{36}$$

$$= \frac{6}{36} + \frac{6}{36} - \frac{1}{36}$$

$$= \frac{12}{36} - \frac{1}{36} = \frac{11}{36}$$

Note that we can't just add $\frac{1}{6}$ and $\frac{1}{6}$ to get $\frac{12}{36}$ or $\frac{1}{3}$, which incidentally is answer choice E. To do so would fail to account for, and properly remove, the overlap created when double sixes are rolled.

After all, what is the probability of getting a six on the first roll of a dice? Answer: (6,1), (6,2), (6,3), (6,4), (6,5), and (6,6). The probability of event A is $\frac{6}{36}$ or $\frac{1}{6}$. What is the probability of getting a six on the second roll of a dice? Answer: (1,6), (2,6), (3,6), (4,6), (5,6), and (6,6). The probability of event B is $\frac{6}{36}$ or $\frac{1}{6}$.

Note that "double sixes" is included in both event A and event B. This overlap must be subtracted out. To be clear, the probability of getting a six on the first or second roll of a dice does include the possibility of getting sixes on the first and second rolls (that is, double sixes), but this outcome

ANSWERS AND EXPLANATIONS

can only be counted once.

Another way this problem could have been asked is: "What is the probability of rolling two normal six-sided dice and getting at least one six?" And yet another way this problem could have been solved is through the use of the Complement Rule of probability. The probability of rolling at least one six is the same as the probability of "one minus the probability of rolling no sixes."

The probability of rolling no sixes is:

$$\frac{5}{6} \times \frac{5}{6} = \frac{25}{36}$$

The probability of rolling at least one six is:

$$1 - \frac{25}{36} = \frac{11}{36}$$

NOTE ↪ Answer choice B would have been the correct answer had the problem asked: "What is the probability of rolling a single dice twice and getting exactly one six?" Perhaps the simplest way to arrive at the correct answer is to write out the possibilities. There are eleven ways to get exactly one six: (6,1), (6,2), (6,3), (6,4),(6,5), (6,6), (1,6), (2,6), (3,6), (4,6), (5,6), and (6,6). Alternatively, we could choose to subtract out $\frac{1}{36}$ from the previous calculation (i.e., $\frac{11}{36}$) in order to remove $\frac{1}{36}$ the probability of rolling double sixes. Note that in the following calculation, the "first" six is removed because it represents overlap while the "second" six is removed because it represents the actual probability of rolling double sixes.

$$\frac{1}{6} + \frac{1}{6} - \frac{1}{36} - \frac{1}{36} = \frac{10}{36} = \frac{5}{18}$$

6. Exam Time Encore

Choice E

29

Overview: This problem involves three overlapping probabilities and, as evident by the "shortcut" approach, it is best solved using the Complement Rule of probability.

I. Shortcut Approach

Using the Complement Rule, we want to calculate the probability of failing all three exams. Then we will subtract this number from 1, in order to determine the probability of passing any and all exams.

i) The probability of not passing the first exam:

$$P(\text{not A}) = 1 - P(A) \quad 1 - \frac{3}{4} = \frac{1}{4}$$

ii) Below is the probability of not passing the second exam:

$$P(\text{not B}) = 1 - P(B) \quad 1 - \frac{2}{3} = \frac{1}{3}$$

iii) Below is the probability of not passing the third exam:

$$P(\text{not C}) = 1 - P(C) \quad 1 - \frac{1}{2} = \frac{1}{2}$$

iv) The probability of failing all three exams:

$$P(\text{not A or B or C}) \quad \frac{1}{4} \times \frac{1}{3} \times \frac{1}{2} = \frac{1}{24}$$

v) The probability of passing at least one exam:

$$P(A) = 1 - P(\text{not A}) \quad 1 - \frac{1}{24} = \frac{23}{24}$$

II. Direct Approach

Using the direct approach, we calculate the probability of passing only one of the three exams, two of the three exams, and all of the three exams. Then, we add these seven results together.

1. Probability of passing exam one but not exams two or three:

ANSWERS AND EXPLANATIONS

$$P(A) \times P(\text{not } B) \times P(\text{not } C) \quad \frac{3}{4} \times \frac{1}{3} \times \frac{1}{2} = \frac{3}{24}$$

2. Probability of passing exam two but not exams one or three:

$$P(\text{not } A) \times P(B) \times P(\text{not } C) \quad \frac{1}{4} \times \frac{2}{3} \times \frac{1}{2} = \frac{2}{24}$$

3. Probability of passing exam three but not exams one or two:

$$P(\text{not } A) \times P(\text{not } B) \times P(C) \quad \frac{1}{4} \times \frac{1}{3} \times \frac{1}{2} = \frac{1}{24}$$

4. Probability of passing exams one and two but not exam three:

$$P(A) \times P(B) \times P(\text{not } C) \quad \frac{3}{4} \times \frac{2}{3} \times \frac{1}{2} = \frac{6}{24}$$

5. Probability of passing exams one and three but not exam two:

$$P(A) \times P(\text{not } B) \times P(C) \quad \frac{3}{4} \times \frac{1}{3} \times \frac{1}{2} = \frac{3}{24}$$

6. Probability of passing exams two and three but not exam one:

$$P(\text{not } A) \times P(B) \times P(C) \quad \frac{1}{4} \times \frac{2}{3} \times \frac{1}{2} = \frac{2}{24}$$

7. Probability of passing all three exams:

$$P(A) \times P(B) \times P(C) \quad \frac{3}{4} \times \frac{2}{3} \times \frac{1}{2} = \frac{6}{24}$$

8. Probability of not passing any of the three exams:

$$P(\text{not } A) \times P(\text{not } B) \times P(\text{not } C) \quad \frac{1}{4} \times \frac{1}{3} \times \frac{1}{2} = \frac{1}{24}$$

These are all the possibilities regarding the outcomes of one student taking three exams. Adding the first seven of eight possibilities above will result in the correct answer using the direct approach.

Proof: $\dfrac{3}{24}+\dfrac{2}{24}+\dfrac{1}{24}+\dfrac{6}{24}+\dfrac{3}{24}+\dfrac{2}{24}+\dfrac{6}{24}=\dfrac{23}{24}$

Note that the total of all eight outcomes above will total to 1 because 1 is the sum total of all probabilistic possibilities.

Proof: $\dfrac{3}{24}+\dfrac{2}{24}+\dfrac{1}{24}+\dfrac{6}{24}+\dfrac{3}{24}+\dfrac{2}{24}+\dfrac{6}{24}+\dfrac{1}{24}=\dfrac{24}{24}=1$

7. Hiring

Choice A

Overview: This particular problem is frequently mistaken for a permutation problem, but does not fall under the umbrella of probability or permutation or combination.

The solution requires only that we multiply together all individual possibilities. Multiplying 7 (that is, candidates for sales managers) by 4 (that is, candidates for shipping clerk) by 10 (that is, candidates for receptionist) would result in 280 possibilities.

$$7\times 4\times 10=280$$

NOTE ✼ This problem is about a series of independent choices. It utilizes the "multiplier principle" and falls within the Rule of Enumeration. The permutation formula is not applicable and cannot be used with this type of problem. This problem is concerned with how many options we have, not how many arrangements are possible, as is the case when dealing with permutation problems.

8. Fencing

Choice C

Overview: This problem is a permutation problem, not a combination problem. In permutation problems, order matters. If country A wins the tournament and country B places second, it is a different outcome than if country B wins and country A places second.

$$P_n^r = \frac{n!}{(n-r)!}$$

$$P_4^2 = \frac{4!}{(4-2)!} = \frac{4!}{2!} = \frac{4 \times 3 \times 2 \times 1}{2 \times 1} = 12$$

NOTE ↤ Consider this follow-up problem. A teacher has four students in a special needs class. She must assign four awards at the end of the year—math, English, history, and creative writing awards. How many ways could she do this assuming that a single student could win multiple awards?

$$n^r = 4^4 \quad 4 \times 4 \times 4 \times 4 = 256$$

She has four ways that she could give out the math award, four ways to give out the English award, four ways to give out the history award, and four ways to give out the creative writing award. Refer to probability formula #10 (page 16).

9. Row

Choice E

Overview: This is a permutation problem which lends itself to the shortcut formula—n!—given that we are utilizing all members of set n, as opposed to a subset of set n.

The problem is essentially asking us: "How many ways can we arrange six people in six seats?" The fact that three of these six individuals are boys and three are girls is irrelevant to the problem at hand. If there are no restrictions on how the students may be seated, there are 720 possibilities with respect to how they can be seated.

$$6! = 6 \times 5 \times 4 \times 3 \times 2 \times 1 = 720$$

There are six ways to seat the first student, five ways to seat the second student, four ways to seat the third student, three ways to seat the fourth student, two ways to seat the fifth student, and only one way to seat the sixth and final student.

10. Alternating

Choice C

Overview: This problem is effectively a joint permutation problem in which we calculate two individual permutations and multiply those outcomes together.

There are two possibilities with respect to how the girls and boys can sit for the make-up exam. Per scenario 1, a boy will sit in the first, third, and fifth seats and a girl will sit in the second, fourth, and sixth seats. Alternatively, per scenario 2, a girl will sit in the first, third, and fifth seats and a boy will sit in the second, fourth, and sixth seats.

Scenario 1:

$$\begin{array}{cccccc} B & G & B & G & B & G \end{array}$$
$$\frac{3}{B_1} \times \frac{3}{G_1} \times \frac{2}{B_2} \times \frac{2}{G_2} \times \frac{1}{B_3} \times \frac{1}{G_3}$$

Scenario 2:

$$\begin{array}{cccccc} G & B & G & B & G & B \end{array}$$
$$\frac{3}{G_1} \times \frac{3}{B_1} \times \frac{2}{G_2} \times \frac{2}{B_2} \times \frac{1}{G_3} \times \frac{1}{B_3}$$

With reference to scenario 1, how many ways can each seat be filled (left to right)? Answer: The first seat can be filled by one of three boys, the second seat can be filled by one of three girls, the third seat can filled by one of two remaining boys, the fourth seat can be filled by one of two remaining girls, the fifth seat will be filled by the final boy, and the sixth seat will be filled by the final girl.

ANSWERS AND EXPLANATIONS

With reference to scenario 2, how many ways can each seat be filled (left to right)? Answer: The first seat can be filled by one of three girls, the second seat can be filled by one of three boys, the third seat can filled by one of two remaining girls, the fourth seat can be filled by one of two remaining boys, the fifth seat will be filled by the final girl, and the sixth seat will be filled by the final boy.

Therefore:

$$B\ G\ B\ G\ B\ G \quad G\ B\ G\ B\ G\ B$$
$$(3\times3\times2\times2\times1\times1)+(3\times3\times2\times2\times1\times1)$$
$$36+36=72$$

In short, the answer can be calculated as follows: $(3!\times3!)+(3!\times3!)$

$$2(3!\times3!)$$
$$2[(3\times2\times1)\times(3\times2\times1)]=72$$
$$2(6\times6)=72$$
$$2(36)=72$$

NOTE ⌇ There is another common variation stemming from this type of permutation problem:

Three boys and three girls are going to sit for a make-up exam. The girls are to sit in the first, second, and third seats while the boys must sit in the fourth, fifth, and sixth seats. How many possibilities are there with respect to how the six students can be seated?

$$\begin{array}{cccccc} B & G & B & G & B & G \\ \dfrac{3}{B_1} \times & \dfrac{3}{G_1} \times & \dfrac{2}{B_2} \times & \dfrac{2}{G_2} \times & \dfrac{1}{B_3} \times & \dfrac{1}{G_3} \end{array}$$

Answer: $3!\times3!=6\times6=36$ possibilities

11. Banana

Choice D

Overview: This problem highlights the handling of "repeated letters" (or "repeated numbers"). The formula for calculating permutations with repeated numbers or letters is $\frac{n!}{x!y!z!}$, where x, y, and z are distinct but identical numbers or letters.

$$\frac{n!}{x!y!} = \frac{6!}{3! \times 2!} = \frac{6 \times 5 \times 4 \times 3 \times 2 \times 1}{3 \times 2 \times 1 \times 2 \times 1} = 60$$

The word "banana" has three "a's" and two "n's"—3! denotes the three "a's" while 2! denotes the two "n's."

12. Table

Choice C

Overview: This problem deals with the prickly issue of "empty seats."

$$\frac{5!}{2!} = \frac{5 \times 4 \times 3 \times 2 \times 1}{2 \times 1} = 60$$

2! represents the two empty seats.

NOTE ✎ The answer to this problem is similar in approach to that of the problem 11, Banana. In permutation theory, "empty seats" are analogous to "identical numbers" (or "identical letters"). Think of the two empty chairs as representing two identical people. Also, the geometric configuration of a round table should not be a distraction. The solution to this problem would be identical had we been dealing with a row of five seats. After all, a table is but a row attached at both its ends and, in the case of a round table, shaped like a circle.

13. Singer

Choice C

Overview: Joint combination problems are solved by multiplying the results of two individual combinations.

ANSWERS AND EXPLANATIONS

First, break the combination into two calculations. First, the "old songs," and second, the "new songs."

Old songs:

$$C_n^r = \frac{n!}{r!(n-r)!}$$

$$C_6^4 = \frac{6!}{4!(6-4)!} = \frac{6!}{4!(2!)} = \frac{6\times 5\times 4\times 3\times 2\times 1}{4\times 3\times 2\times 1\times (2\times 1)} = 15$$

Thus, 15 represents the number of ways the singer can choose to sing four of six old songs.

New songs:

$$C_n^r = \frac{n!}{r!(n-r)!}$$

$$C_5^2 = \frac{5!}{2!(5-2)!} = \frac{5!}{2!(3!)} = \frac{5\times 4\times 3\times 2\times 1}{2\times 1\times (3\times 2\times 1)} = 10$$

Thus, 10 represents the number of ways the singer could chose to sing three of five new songs. Therefore, the joint combination equals $15\times 10=150$.

In summary, the outcome of this joint combination is:

$$C_n^r \times C_n^r = \frac{n!}{r!(n-r)!} \times \frac{n!}{r!(n-r)!}$$

$$C_6^4 \times C_5^2 = \frac{6!}{4!(6-4)!} \times \frac{5!}{2!(5-2)!}$$

$$C_6^4 \times C_5^2 = \frac{6!}{4!(2)!} \times \frac{5!}{2!(3)!}$$

$$C_6^4 \times C_5^2 = \frac{6\times 5\times 4!}{4!(2!)} \times \frac{5\times 4\times 3!}{2!(3!)} = 15\times 10 = 150$$

14. Reunion

Choice D

Overview: This problem appears rather complicated on the surface, but its solution is actually quite simple. We're essentially asking: "How many groups of two can we create from a group of eleven where order doesn't matter? Or specifically: "How many ways can we choose two people from eleven people where order doesn't matter?"

$$C_n^r = \frac{n!}{r!(n-r)!}$$

$$C_{11}^2 = \frac{11!}{2!(11-2)!} = \frac{11 \times 10 \times 9!}{2(9!)} = 55$$

NOTE ⌘ This problem would have resulted in the same answer had it asked: "How many teams of two can we create from 11 persons in order to stage a beach volley ball tournament?" The formation of teams is a classic example of combinations in that order doesn't matter. It doesn't matter which member of a team is chosen first or second, a team of two is simply a team.

15. Outcomes

Choice A

Overview: This problem exists to test permutation and combination theory at a grass roots level. A strong understanding of theory will eliminate the need to do any calculations.

Statement I:

True. $P_5^3 > P_5^2$

$P_5^3 = 60$ and $P_5^2 = 20$. Order matters in permutations and the more items there are in a permutation the more possibilities there are.

Statement II:

False. $C_5^3 > C_5^2$

ANSWERS AND EXPLANATIONS

$C_5^3 = 10$ and $C_5^2 = 10$. Strange as it may seem, the outcomes are equal! "Complements in combinations" result in the same number of outcomes. Complements occur when the two inside numbers equal the same outside number. Here 3+2=5. Note that this phenomenon occurs only in combinations and not in permutations.

Statement III:

False. $C_5^2 > P_5^2$

$C_5^2 = 10$ and $P_5^2 = 20$. Order matters in permutations and this creates more possibilities relative to combinations. Stated in the reverse, order doesn't matter in combinations and this results in fewer outcomes than permutations, all things being equal.

About the Author

Brandon Royal (CPA, MBA) is an award-winning writer whose educational authorship includes The Little Green Math Book, The Little Blue Reasoning Book, The Little Red Writing Book, and The Little Gold Grammar Book. During his tenure working in Hong Kong for US-based Kaplan Educational Centers—a Washington Post subsidiary and the largest test-preparation organization in the world—Brandon honed his theories of teaching and education and developed a set of key learning "principles" to help define the basics of writing, grammar, math, and reasoning.

A Canadian by birth and graduate of the University of Chicago's Booth School of Business, his interest in writing began after completing writing courses at Harvard University. Since then he has authored a dozen books and reviews of his books have appeared in Time Asia magazine, Publishers Weekly, Library Journal of America, Midwest Book Review, The Asian Review of Books, Choice Reviews Online, Asia Times Online, and About.com. Brandon is a five-time winner of the International Book Awards, a five-time gold medalist at the President's Book Awards, as well as a winner of the Global eBook Awards, the USA Book News "Best Book Awards," and recipient of the 2011 "Educational Book of the Year" award as presented by the Book Publishers Association of Alberta.

Books by Brandon Royal

The Little Blue Reasoning Book:
  50 *Powerful Principles for Clear and Effective Thinking*

The Little Red Writing Book:
  20 *Powerful Principles for Clear and Effective Writing*

The Little Gold Grammar Book:
  Mastering the Rules That Unlock the Power of Writing The Little Red Writing Book Deluxe Edition

The Little Green Math Book:
  30 *Powerful Principles for Building Math and Numeracy Skills*

The Little Purple Probability Book:
  Mastering the Thinking Skills That Unlock the Secrets of Basic Probability

Game Plan for Getting into Business School Game Plan for the GMAT Game Plan for GMAT Math Game Plan for GMAT Verbal

Dancing for Your Life:
  The True Story of Maria de la Torre and Her Secret Life in a Hong Kong Go-Go Bar

The Map Maker:
  An Illustrated Short Story About How Each of Us Sees the World Differently and Why Objectivity is Just an Illusion

Paradise Island:
  An Armchair Philosopher's Guide to Human Nature (or "Life Lessons You Learn While Surviving Paradise")